HARCOURT SCHOOL PUBLISHERS

Practice Book

 Developed by Education Development Center, Inc. through National Science Foundation Grant No. ESI-0099093

Visit *The Learning Site!*
www.harcourtschool.com/thinkmath

Copyright © by Education Development Center, Inc.

All rights reserved. No part of this publication may be reproduced or transmitted in any form or by any means, electronic or mechanical, including photocopy, recording, or any information storage and retrieval system, without permission in writing from the publisher.

Permission is hereby granted to individuals using the corresponding student's textbook or kit as the major vehicle for regular classroom instruction to photocopy entire pages from this publication in classroom quantities for instructional use and not for resale. Requests for information on other matters regarding duplication of this work should be addressed to School Permissions and Copyrights, Harcourt, Inc., 6277 Sea Harbor Drive, Orlando, Florida 32887-6777. Fax: 407-345-2418.

HARCOURT and the Harcourt Logo are trademarks of Harcourt, Inc., registered in the United States of America and/or other jurisdictions.

Printed in the United States of America

ISBN 13: 978-0-15-342493-9

ISBN 10: 0-15-342493-1

3 4 5 6 7 8 9 10 170 16 15 14 13 12 11 10 09 08

If you have received these materials as examination copies free of charge, Harcourt School Publishers retains title to the materials and they may not be resold. Resale of examination copies is strictly prohibited and is illegal.

Possession of this publication in print format does not entitle users to convert this publication, or any portion of it, into electronic format.

This program was funded in part through the National Science Foundation under Grant No. ESI-0099093. Any opinions, findings, and conclusions or recommendations expressed in this program are those of the authors and do not necessarily reflect the views of the National Science Foundation.

Think Math! Contents

Chapter 1 Two-Dimensional Figures and Patterns

1.1 Introducing *Think Math!* **P1**
1.2 Examining Two-Dimensional Figures... **P2**
1.3 Sorting by Attributes **P3**
1.4 Sorting by More Than One Attribute ... **P4**
1.5 Counting Differences **P5**
1.6 Repeating and Growing Patterns **P6**

Chapter 2 Number Lines and Time

2.1 Introducing the Number Line **P7**
2.2 Jumping on the Number Line **P8**
2.3 Recording Jumps as Addition and Subtraction **P9**
2.4 Relating Addition and Subtraction **P10**
2.5 Comparing Numbers on the Number Line **P11**
2.6 Comparing Numbers and Quantities **P12**
2.7 Investigating Time and Events **P13**
2.8 Telling Time to the Hour **P14**
2.9 Ordinal Numbers and the Calendar... **P15**

Chapter 3 Skip-Counting and Money

3.1 Introducing the Penny **P16**
3.2 Counting with Groups **P17**
3.3 Introducing the Nickel **P18**
3.4 Counting Money **P19**
3.5 Making Amounts in Different Ways **P20**

Chapter 4 Exploring Addition and Subtraction

4.1 Introducing the Cross Number Puzzle **P21**
4.2 Using Cross Number Puzzles to Add **P22**
4.3 Exploring Missing Addends **P23**
4.4 Practice with Cross Number Puzzles **P24**
4.5 Sums of 10 .... **P25**
4.6 Addition Stories **P26**
4.7 Subtraction Stories **P27**

Chapter 5 Working with Tens

5.1 Ten and Some More **P28**
5.2 Lots of Tens and Some More **P29**
5.3 Using Dimes and Pennies................... **P30**
5.4 Tens and Time **P31**
5.5 Tens on the Number Line **P32**
5.6 Using the Number Line to Solve Problems **P33**
5.7 Modeling Numbers in Different Ways **P34**

Chapter 6 Data and Probability

6.1 Collecting and Tallying Data.............. **P35**
6.2 Making Graphs with Objects and Pictures **P36**
6.3 Making Graphs with Pictures and Symbols **P37**
6.4 Bar Graphs and Probability **P38**
6.5 Investigating Probability **P39**

Contents

Chapter 7 Working with Larger Numbers

7.1	Identifying Rules	P40
7.2	Identifying Rules with Larger Numbers	P41
7.3	Adding Ten on the Number Line Hotel	P42
7.4	Subtracting Ten on the Number Line Hotel	P43
7.5	Adding and Subtracting with Larger Numbers	P44
7.6	Modeling Numbers to 99	P45
7.7	Numbers to 100 and Beyond	P46
7.8	Connecting Numbers and Words	P47
7.9	Introducing the Quarter	P48

Chapter 8 Doubling, Halving, and Fractions

8.1	Mirrors and Folding	P49
8.2	Doubling Your Money	P50
8.3	Sharing to Find Half	P51
8.4	Exploring One Half	P52
8.5	Wholes and Halves	P53
8.6	Halfway Between Whole Numbers	P54
8.7	Half of a Half	P55
8.8	Thirds	P56

Chapter 9 Modeling Addition and Subtraction

9.1	Exploring Addition with Cuisenaire® Rods	P57
9.2	Recording Addition Sentences	P58
9.3	Exploring Input/Output Tables	P59
9.4	Using Input/Output Tables	P60
9.5	Making Fact Families	P61
9.6	Fact Families and Stair-Step Numbers	P62
9.7	Connecting Stories and Fact Families	P63
9.8	Two-Sentence Fact Families	P64

Chapter 10 Maps, Grids, and Geometric Figures

10.1	Exploring Lines and Intersections	P65
10.2	Drawing Lines and Intersections	P66
10.3	Exploring Direction on a Map	P67
10.4	Finding and Following Paths on a Grid	P68
10.5	Paths and Figures on a Grid	P69
10.6	Exploring Symmetry	P70
10.7	Connecting Points to Make Figures	P71
10.8	Investigating Rectangles	P72
10.9	Recording and Graphing Rectangles	P73
10.10	Finding Congruent Figures	P74
10.11	Exploring Three-Dimensional Figures	P75

Chapter 11 Comparing Numbers, Temperatures, and Weights

11.1	Comparing Groups	**P76**
11.2	Comparing Numbers and Temperatures	**P77**
11.3	Using Place Value to Compare Numbers	**P78**
11.4	Ordering Numbers	**P79**
11.5	Changing Both Sides of a Sentence	**P80**
11.6	Comparing and Ordering Weights	**P81**
11.7	Changing Both Pans of a Balance	**P82**

Chapter 12 Length, Area, and Capacity

12.1	Measuring Length with Nonstandard Units	**P83**
12.2	Comparing and Ordering Lengths	**P84**
12.3	Measuring with a Centimeter Ruler	**P85**
12.4	Measuring with an Inch Ruler	**P86**
12.5	Comparing Figures by Size	**P87**
12.6	Exploring Area	**P88**
12.7	Finding Area on a Grid	**P89**
12.8	Comparing Areas	**P90**
12.9	Measuring Boxes and Rectangles	**P91**
12.10	Introducing Capacity with Nonstandard Units	**P92**
12.11	Measuring Capacity with Standard Units	**P93**

Chapter 13 Making and Breaking Numbers

13.1	Making Even and Odd Numbers	**P94**
13.2	Making Numbers as Sums of 1, 2, 4, and 8	**P95**
13.3	Combining Triangular Numbers	**P96**
13.4	Making Sums of 60	**P97**
13.5	Sums to 12	**P98**
13.6	Sums to 15	**P99**
13.7	Sums to 16	**P100**
13.8	Sums to 18	**P101**
13.9	Sums to 20	**P102**

Chapter 14 Extending Addition and Subtraction

14.1	Adding Number Sentences	**P103**
14.2	Making Addition Easier	**P104**
14.3	Modeling Number Sentences and Stories	**P105**
14.4	Making Subtraction Easier	**P106**
14.5	Subtraction That Changes the Tens Digit	**P107**
14.6	Change from a Dollar	**P108**
14.7	Is This Story Reasonable?	**P109**
14.8	Solving Puzzles with Many Solutions	**P110**

Contents

Chapter 15 Exploring Rules and Patterns

15.1　Identifying Rules **P111**

15.2　Sorting Rules **P112**

15.3　Undoing Rules **P113**

15.4　Rules with More Than One Input **P114**

15.5　Conversion Rules **P115**

15.6　Skip-Counting with Money **P116**

15.7　Creating Figures and Patterns **P117**

15.8　Patterns with Skip-Counting **P118**

15.9　Relating One-Color Trains **P119**

These pages provide additional practice for each lesson in the chapter. The exercises are used to reinforce the skills being taught in each lesson.

Practice Book

Name _____ Date _____

Practice
Lesson 1

● Introducing *Think Math!*

What numbers are missing?

 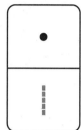 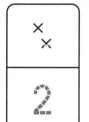  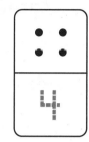

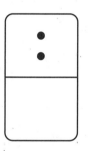 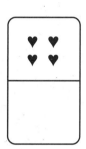 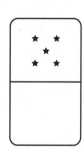

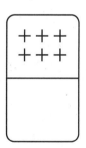 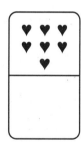

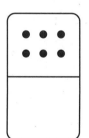

Chapter 1 Practice Book **P1**

Name _____ Date _____

Examining Two-Dimensional Figures

Color ▭ green.   Color ○ blue.

Color △ yellow.   Draw an X on □.

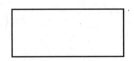

      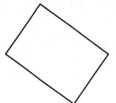

Name _____ Date _____

Practice
Lesson 3

● Sorting by Attributes

Cross out the figure that does not belong.

1.

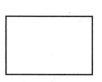  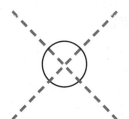 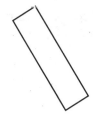

2.

  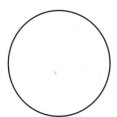

3.

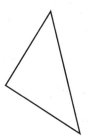  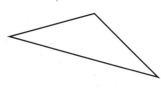

4.

 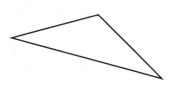 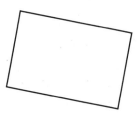

Chapter 1

Name _____ Date _____

Practice
Lesson 4

Sorting by More Than One Attribute

Draw each figure.

1. large blue square

2. small yellow triangle

3. small red circle

4. large blue circle

5. small yellow rectangle

6. large red triangle

P4 Practice Book

Chapter 1

Counting Differences

Circle the figure that is different in one way.

1.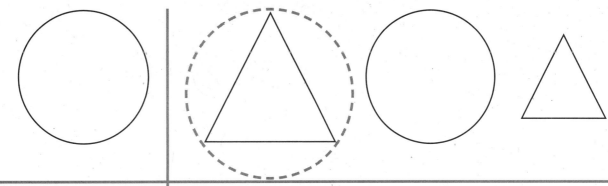

2.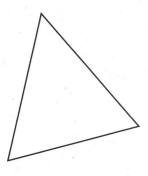

Circle the figure that is different in two ways.

3.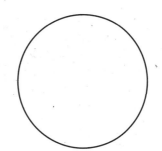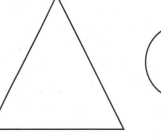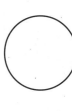

4.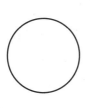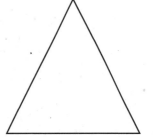

Name _____ Date _____

Practice
Lesson 6

Repeating and Growing Patterns

Which figure is missing?

1.

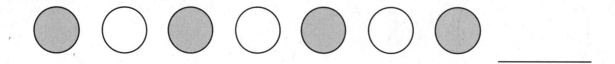

2.

3.

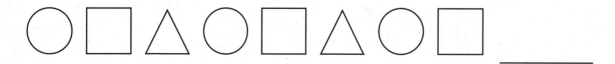

4.

P6 Practice Book   Chapter 1

Name _____ Date _____

Practice
Lesson 1

Introducing the Number Line

Continue each pattern.

1.

2.

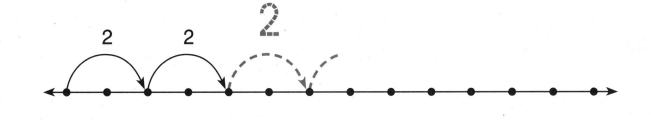

3.

4.

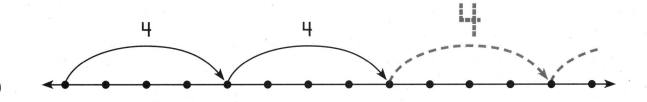

Chapter 2 Practice Book **P7**

Name _____ Date _____

Practice Lesson 2

Jumping on the Number Line

Continue each pattern.

1.

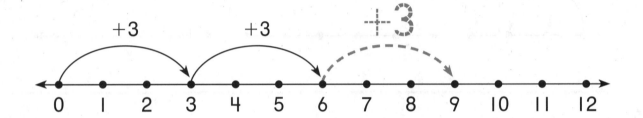

2.

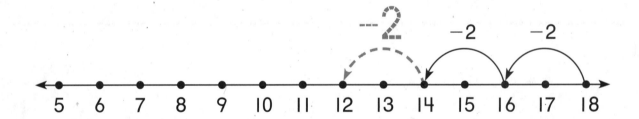

3.

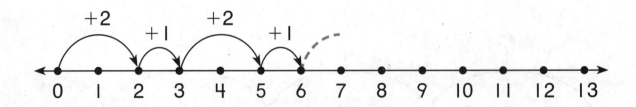

P8 Practice Book Chapter 2

Name _____ Date _____

**Practice
Lesson 3**

● Recording Jumps as Addition and Subtraction

Draw the jumps.
Find the missing numbers.

1.

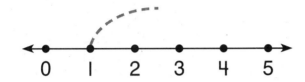

2.

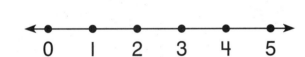

3.

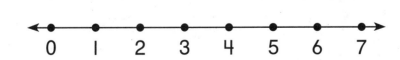

Chapter 2

Name _____ Date _____

Practice
Lesson 4

Relating Addition and Subtraction

Draw the jumps.
Find the missing numbers.

1.
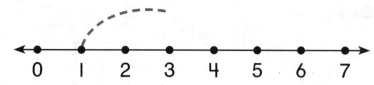

2.
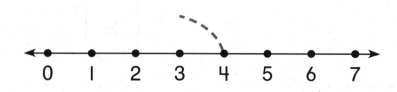

3.
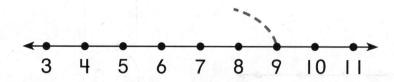 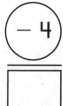

P10 Practice Book Chapter 2

Practice
Lesson 5

• Comparing Numbers on the Number Line

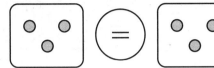 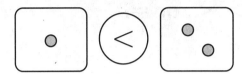

Write >, <, or =.

1.

2.

3.

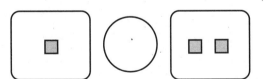

4.

5.

6.

Chapter 2

Name _____ Date _____

Practice
Lesson 6

Comparing Numbers and Quantities

Write >, <, or =.

1.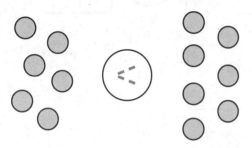

   6   7

2.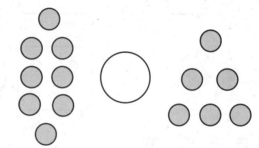

3.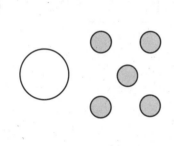

4.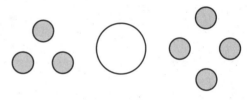

5.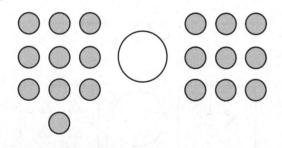

6.

Name _____ Date _____

Practice
Lesson 7

Investigating Time and Events

1. Draw something you do in the morning.

2. Draw something you do in the evening.

3. Order the events from least amount of time to most amount of time. Write 1, 2, and 3 to show the order.

Sleep

Brush hair

Eat lunch

_____   _____   _____

Chapter 2   Practice Book **P13**

Name _____ Date _____

Practice Lesson 8

Telling Time to the Hour

What time is it?

1.

2.

3.

4.

5.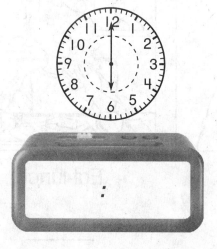

6.

Name _____ Date _____

Practice Lesson 9

Ordinal Numbers and the Calendar

November

Sunday	Monday	Tuesday	Wednesday	Thursday	Friday	Saturday
			1	2	3	4
5	6	7	8	9	10	11
12	13	14	15	16	17	18
19	20	21	22	23	24	25
26	27	28	29	30		

1. Color the third Wednesday orange.
2. Color the fourth Tuesday yellow.
3. Color the sixth day of the month green.
4. What day is the tenth day of the month?

   _____

5. What day is the second day of the month?

   _____

Chapter 2   Practice Book **P15**

Name _____ Date _____

Practice
Lesson 1

Introducing the Penny

What is the value?

1.
_5_ ¢

2.
_____ ¢

3.
_____ ¢

4.
_____ ¢

5.
_____ ¢

6.
_____ ¢

Name _____ Date _____

Practice
Lesson 2

● Counting with Groups

How many are there?

1.   _15_

2.   _____

3.   _____

4. 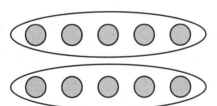   _____

5. 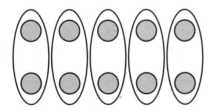   _____

6. 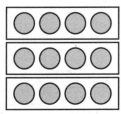   _____

Chapter 3

Name _____ Date _____

Practice
Lesson 3

Introducing the Nickel

 is a nickel, worth 5¢.   is a penny, worth 1¢.

What is the total value?

1. = 6 ¢

2. = ___ ¢

3. = ___ ¢

4. +  = ___ ¢

5. +  = ___ ¢

6.  + = ___ ¢

P18 Practice Book   Chapter 3

Name _____ Date _____

Practice
Lesson 4

Counting Money

 is a penny, worth 1¢.   is a nickel, worth 5¢.

How many coins are there?
What is the value?

1. P

    _____ coin

    _____ ¢

2. N

    _____ coin

    _____ ¢

3.

    _____ coins

    _____ ¢

4.

    _____ coins

    _____ ¢

5.

    _____ coins

    _____ ¢

6.

    _____ coins

    _____ ¢

7.

    _____ coins

    _____ ¢

8.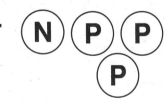

    _____ coins

    _____ ¢

9.

    _____ coins

    _____ ¢

Chapter 3

Practice Book **P19**

Name _____ Date _____

Practice Lesson 5

Making Amounts in Different Ways

 is a nickel, worth 5¢.   is a penny, worth 1¢.

Color to show the amount.

1.   +  = 8¢

2.   +  = 13¢

3.   +     = 6¢

4.   +     = 11¢

5.   +    = 9¢

Name _____ Date _____

Practice Lesson 1

Introducing the Cross Number Puzzle

What numbers are missing?

1.

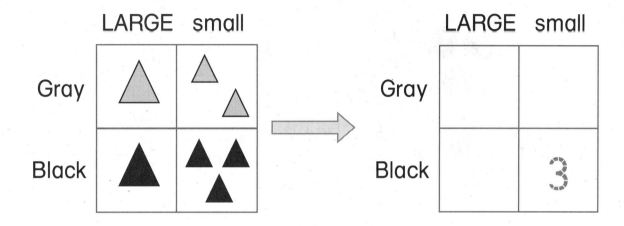

2.

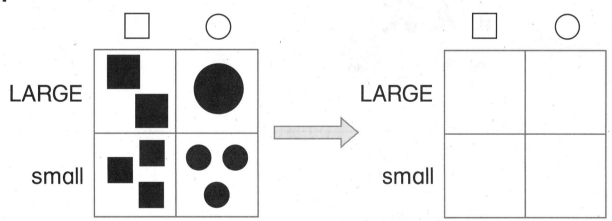

Chapter 4 — Practice Book **P21**

Name _____ Date _____

Practice
Lesson 2

Using Cross Number Puzzles to Add

Write a number sentence for each row.

1.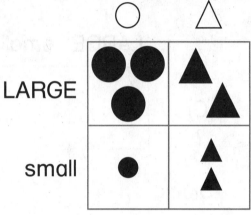

   ___3___ + ___2___ = _____

   _____ + _____ = _____

2.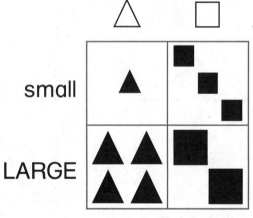

   _____ + _____ = _____

   _____ + _____ = _____

3.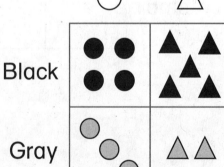

   _____ + _____ = _____

   _____ + _____ = _____

P22 Practice Book

Chapter 4

Practice
Lesson 3

• Exploring Missing Addends

What numbers are missing?

1.

4	1	5
0		1
4		

4 + 1 = 5

0 + ☐ = 1

$$\begin{array}{r} 4 \\ + \, 0 \\ \hline 4 \end{array} \quad \begin{array}{r} 1 \\ + \, \square \\ \hline \square \end{array}$$

2.

2		4
	4	4
2	6	

2 + ☐ = 4

☐ + 4 = 4

$$\begin{array}{r} 2 \\ + \, \square \\ \hline 2 \end{array} \quad \begin{array}{r} \square \\ + \, 4 \\ \hline 6 \end{array}$$

3.

	1	3
3		4

☐ + 1 = 3

3 + ☐ = 4

$$\begin{array}{r} \square \\ + \, 3 \\ \hline \square \end{array} \quad \begin{array}{r} 1 \\ + \, \square \\ \hline \square \end{array}$$

Chapter 4
Practice Book **P23**

Name _____ Date _____

Practice
Lesson 4

Practice with Cross Number Puzzles

What numbers are missing?

1.

0	1	
4		7

2.

2	0	
2		
	4	

3.

0		3
	1	
1		

4.

	3	5
2		
	4	

5.

	3	
		1
2		6

6.

		2
4		
	4	9

P24 Practice Book   Chapter 4

Name _____ Date _____

Practice
Lesson 5

Sums of 10

Complete each puzzle. Use the puzzle to complete the number sentence.

1.

1	2	3
4	3	
5		

5 + ____ = 3 + ____

2.

5	1	
	3	10

____ + 3 = ____ + ____

3.

0		
	1	
8		10

8 + ____ = ____ + ____

Chapter 4   Practice Book **P25**

Name _____ Date _____

Practice
Lesson 6

Addition Stories

Solve each problem. Show your work with pictures, numbers, or words.

1. Mia drew 4 circles and 2 squares.
   How many shapes did she draw?

   _____ shapes

2. 5 children were at a party.
   Then 3 more joined them.
   How many children are at the party now?

   _____ children

3. All fish in a tank are blue or gold.
   7 fish are blue. 3 fish are gold.
   How many fish are in the tank?

   _____ fish

Name _____ Date _____

Practice
Lesson 7

Subtraction Stories

Solve each problem. Show your work with pictures, numbers, or words.

1. There are 6 ducks in a pond. 3 swim away. How many ducks are left?

   _____ ducks

2. Rita has 4 sweaters.
   2 have zippers.
   The rest have buttons.
   How many sweaters have buttons?

   _____ sweaters

3. Lou takes out 9 books from the library.
   Kim takes out 4 books.
   How many more books did Lou take out than Kim?

   _____ more books

Chapter 4

Ten and Some More

What numbers are missing?

1.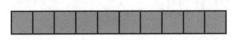

 $\begin{array}{r} 10 \\ +2 \\ \hline 12 \end{array}$

2.

 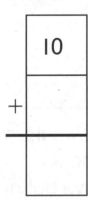
 $\begin{array}{r} 10 \\ + \\ \hline \end{array}$

3.

 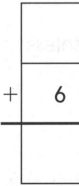
 $\begin{array}{r} \\ +6 \\ \hline \end{array}$

4.

 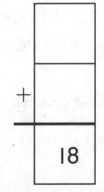
 $\begin{array}{r} \\ + \\ \hline 18 \end{array}$

Name _____ Date _____

Practice
Lesson 2

Lots of Tens and Some More

Write the missing numbers.

1.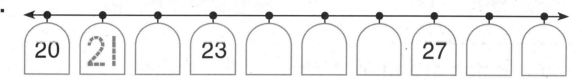
20, 21, ___, 23, ___, ___, ___, 27, ___, ___

2. $20 + 9 = 29$

3. $\square + 3 = 23$

4. $\square + 8 = 28$

5. $20 + \square = 21$

6.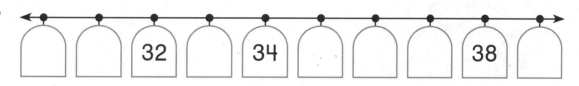
___, ___, 32, ___, 34, ___, ___, ___, 38, ___

7. $\square + 7 = 37$

8. $30 + 3 = \square$

9. $30 + \square = 34$

10. $\square + 6 = 36$

Chapter 5

Practice Book **P29**

Name _____ Date _____

Practice
Lesson 3

Using Dimes and Pennies

 is a dime, worth 10¢.   is a penny, worth 1¢.

How many coins are there? What is the value?

1.

   __6__ coins

   _____ ¢

2.

   _____ coins

   _____ ¢

3.

   _____ coins

   _____ ¢

4.

   _____ coins

   _____ ¢

5. What numbers are missing?

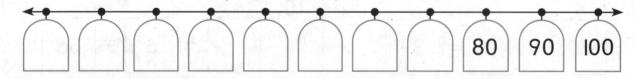

Name _____ Date _____

Practice
Lesson 4

Tens and Time

What time is it?

1.

   2:00   ___:___

2.

   ___:___   ___:___

3.

   ___:___   ___:___

4.

   ___:___   ___:___

Chapter 5

Tens on the Number Line

Write the numbers to match each jump.

1.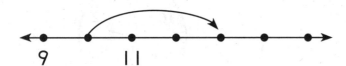

2.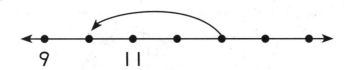

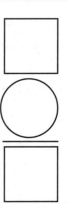

Draw the jump. Complete the number sentence.

3.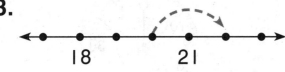

   20 + ☐ = 22

4.

   30 + ☐ = 35

5.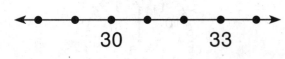

   ☐ + 3 = 33

6.

   ☐ + ☐ = 43

P32 Practice Book   Chapter 5

Name _____ Date _____

Practice
Lesson 6

Using the Number Line to Solve Problems

What numbers match the jump?

1.
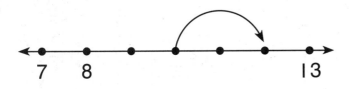

2.
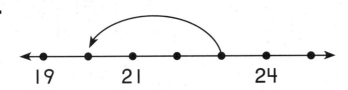

3.

4.
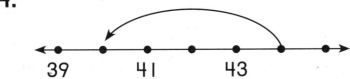

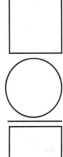

Chapter 5

Name _____ Date _____

Practice Lesson 7

Modeling Numbers in Different Ways

What is the number?

1.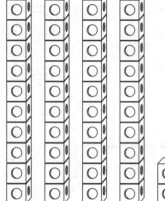

_____

2.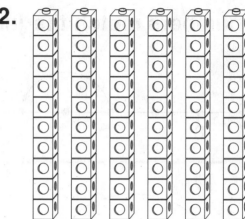

_____

3.

_____

4.

_____

5. 10 + 10 + 6

_____

6. 20 + 10 + 10 + 2

_____

7.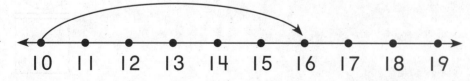

_____

P34 Practice Book Chapter 5

Practice
Lesson 1

Collecting and Tallying Data

Ana asked her classmates if they bring lunch to school.

Do you bring lunch to school?		
no	yes	no
no	no	no
yes	yes	no
yes	yes	yes
no	no	yes
yes	yes	

1. Use tally marks to show the data.

Yes	No

2. How many classmates bring lunch to school?

    _____ classmates

3. How many classmates do not bring lunch to school?

    _____ classmates

4. How many classmates did Ana ask?

    _____ classmates

5. How many more classmates bring lunch than do not bring lunch?

    _____ more classmate

Chapter 6

Practice Book **P35**

Name _____ Date _____

Practice Lesson 2

Making Graphs with Objects and Pictures

Jody saw some animals on a nature hike.

Animals Jody Saw					
deer	🦌	🦌			
raccoons	🦝	🦝	🦝		
birds	🐦	🐦	🐦	🐦	🐦
coyotes	🐺				

1. She saw 2 _____.

2. _____ are the animals she saw the most.

3. She saw ____ raccoons.

4. She saw ____ animals in all.

Shawn sorts his toy cars by size.

5. There are ____ small cars.

6. The size with the fewest cars is _____.

7. There are more _____ cars than _____ cars.

8. There are ____ cars in all.

Sizes of Cars

(small: 7, medium: 3, large: 6)

small medium large

P36 Practice Book

Chapter 6

Practice
Lesson 3

Making Graphs with Pictures and Symbols

Some children chose their favorite season. The table shows the results.

Our Favorite Seasons	
summer	ℍℍ II
fall	III
winter	ℍℍ
spring	IIII

1. Use the data in the table to complete the graph.

Our Favorite Seasons								
summer								
fall								
winter								
spring								

Key: Each ☺ stands for 1 child's choice.

2. Which season did most children choose? _____

3. How many more children chose summer than winter?

   _____ more children

4. How many children chose fall or winter? Explain how you know.

   _____

   _____

Chapter 6   Practice Book **P37**

Name _____ Date _____

Practice
Lesson 4

Bar Graphs and Probability

Brooke and Evan played the Number Race game.

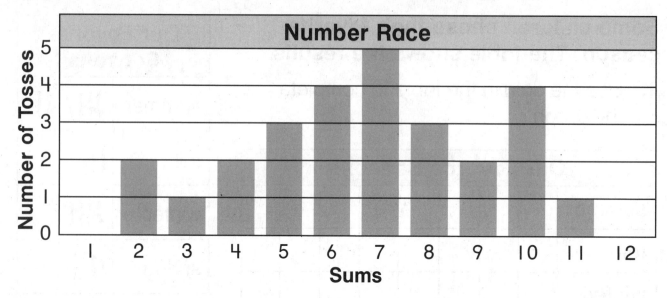

1. What is the winning number? _____

2. How many more times did Brooke and Evan toss the number 7 than the number 3?

   _____ more times

3. How many times did Brooke and Evan toss

   the number cubes? _____ times

4. How did you find the answer to Problem 3?

   _____

   _____

P38 Practice Book   Chapter 6

Name _____ Date _____

Practice
Lesson 5

Investigating Probability

Is it *possible* or *impossible*?
Draw lines to match.

1. It will rain tomorrow. ---------- possible

2. The school lunch will have milk.

   impossible

3. A cow will jump over the moon.

4. Draw a picture to show an *impossible* event.

Is it *certain*, *likely*, or *unlikely*?
Draw lines to match.

5. A penny is worth 1¢.   certain

6. One apple costs more than 3 apples.   likely

7. Three apples cost more than 1 apple.   unlikely

Chapter 6   Practice Book   **P39**

Name _____ Date _____

Practice
Lesson 1

Identifying Rules

What is the rule?

1.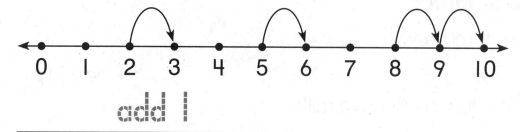

   _____ add 1 _____

2.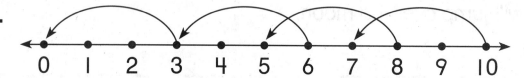

   _____

Draw the missing jumps. Complete each table.

3.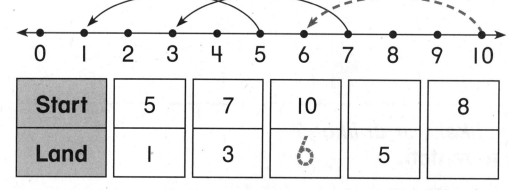

Start	5	7	10		8
Land	1	3	6	5	

4.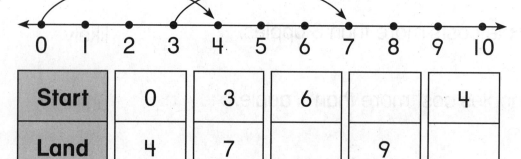

Start	0	3	6		4
Land	4	7		9	

P40 Practice Book   Chapter 7

Name _____ Date _____

Practice Lesson 2

Identifying Rules with Larger Numbers

Draw the missing jumps.
Complete each table.

1.

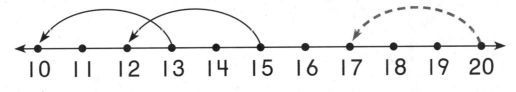

Start	13	15	20	17	
Land	10	12	17		16

2.

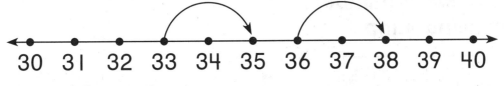

Start	33	36	30		38
Land	35	38		37	

3.

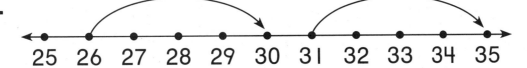

Start	26	31	27		30
Land	30	35		33	

Chapter 7   Practice Book **P41**

Name _____ Date _____

Practice Lesson 3

Adding 10 on the Number Line Hotel

**What is the missing number?
Draw the missing jump.**

1. 10 + 10 = 20

2. 24 + 10 = ☐

3. 2 + 10 = ☐

4. 18 + 10 = ☐

5. 33 + 10 = ☐

6. 5 + 10 = 15

7. ☐ + 10 = 40

8. ☐ + 10 = 25

9. ☐ + 10 = 32

10. ☐ + 10 = 23

P42 Practice Book   Chapter 7

Name _____ Date _____

Practice Lesson 4

Subtracting 10 on the Number Line Hotel

What is the missing number? Draw the missing jump.

1. 16 − 10 = 6
2. 40 − 10 = ☐
3. 14 − 10 = ☐
4. 23 − 10 = ☐
5. 33 − 10 = ☐
6. 12 − 10 = 2
7. ☐ − 10 = 15
8. ☐ − 10 = 11
9. ☐ − 10 = 34
10. ☐ − 10 = 8

Chapter 7

Adding and Subtracting with Larger Numbers

Use the Number Line Hotel to help.

What numbers are missing?

1.
Start	10	27	30	41	18		
Jump Forward	3	1	9	4	2	10	7
Land	13	28	39			29	10

2.
Start	15	29	17	38		40	
Jump Back	5	3	2	8	1	5	10
Land	10	26			23		64

3.
Start	13	26	14	16			39
Jump Back	2	6	4	1	7	10	5
Land	11	20			40	12	

Practice
Lesson 6

Name _____ Date _____

Modeling Numbers to 99

What is the number?

1. 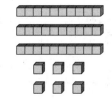 12

2. 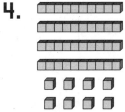 ☐

3. 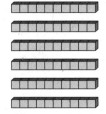 ☐

4. 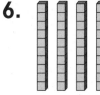 ☐

5. ☐

6. ☐

7. 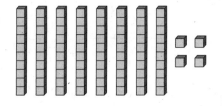 ☐

8. 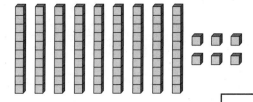 ☐

Chapter 7
Practice Book **P45**

Practice
Lesson 7

Numbers to 100 and Beyond

What numbers are missing?

1.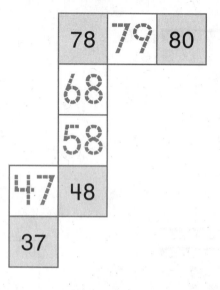

2.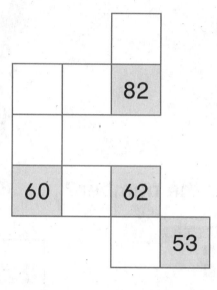

3.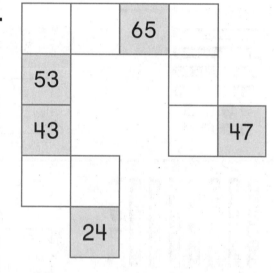

4.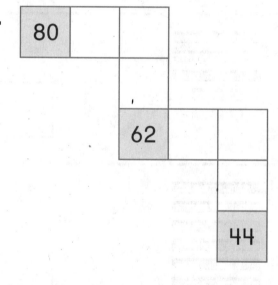

5.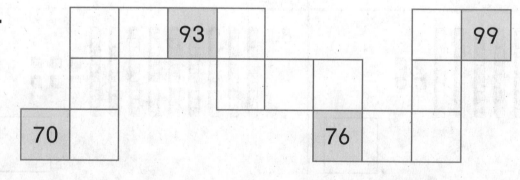

P46 Practice Book

Chapter 7

Name _____ Date _____

Practice
Lesson 8

Connecting Numbers and Words

Write each number.

1. thirteen	2. eighteen
3. forty-three	4. twenty-six
5. thirty-nine	6. sixty-two
7. fifty-seven	8. seventy-five
9. ninety-one	10. twelve

Write each word name.

11. 57

12. 25 _____

13. 32 _____

14. 53 _____

Chapter 7   Practice Book **P47**

Practice
Lesson 9

Introducing the Quarter

What is the value?

- Q is a quarter.
- D is a dime.
- N is a nickel.
- P is a penny.

1.  _____ ¢

2. 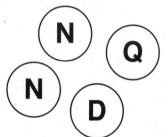 _____ ¢

3. 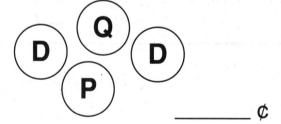 _____ ¢

4. 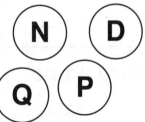 _____ ¢

5.  _____ ¢

6. 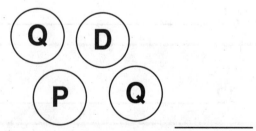 _____ ¢

Name _____ Date _____

Practice
Lesson 1

Mirrors and Folding

Draw the mirror image.

1.

2.

3.

4.

Draw a line to show halves.

 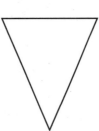 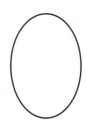

  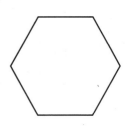

Chapter 8

Practice Book **P49**

Practice Lesson 2

Doubling Your Money

Match. Complete each addition sentence.

1. If one sticker costs 8¢, then how much do two stickers cost?

2. If one sticker costs 12¢, then how much do two stickers cost?

3. If two stickers cost 12¢, then how much does one sticker cost?

4. If two stickers cost 24¢, then how much does one sticker cost?

5. If one sticker costs 6¢, then how much do two stickers cost?

12¢ + 12¢ = ☐

☐ + ☐ = 24¢

6¢ + 6¢ = ☐

8¢ + 8¢ = 16¢

☐ + ☐ = 12¢

P50 Practice Book — Chapter 8

Practice
Lesson 3

Name _____ Date _____

Sharing to Find Half

**Complete each sentence.
Use counters to help.**

1. Two friends share 10 crayons equally.

   Each friend gets ___5___ crayons

2. _____ friends share 16 crayons equally.

   Each friend gets 8 crayons.

3. Two friends share 30 crayons equally.

   Each friend gets _____ crayons.

4. Two friends share _____ crayons equally.

   Each friend gets 25 crayons.

5. Two friends share 26 crayons equally.

   Each friend gets _____ crayons.

Chapter 8

Name _____ Date _____

Practice
Lesson 4

Exploring One Half

Trina is half as old as Tracy. Complete the table.

	Trina	Tracy
1.	1 year	2 years
2.	____ years	10 years
3.	9 years	____ years
4.	____ years	26 years
5.	15 years	____ years
6.	____ years	42 years
7.	25 years	____ years

Name _____ Date _____

Practice
Lesson 5

Wholes and Halves

How many apples are there?

1.

    ___0___ 🍎

2.

    _____ 🍎

3.

    _____ 🍎

4.

    _____ 🍎

5.

    _____ 🍎

6.

    _____ 🍎

7.

    _____

8.

    _____ 🍎

9. Write the numbers above in order from smallest to largest.

    __0__ ____ ____ ____ ____ ____ ____ ____

Chapter 8   Practice Book **P53**

Name _____ Date _____

Practice Lesson 6

Halfway Between Whole Numbers

Complete the number line.
Then circle the number in the middle.

1.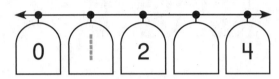

2.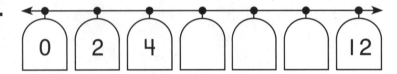

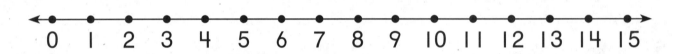

What number is in the middle?
Use the number line to help.

3.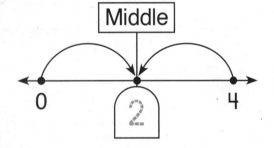

4.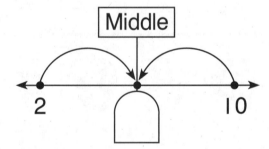

5.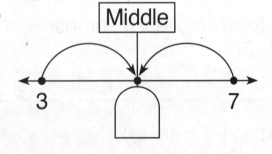

6.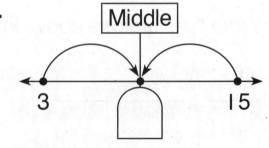

P54 Practice Book   Chapter 8

Name _____ Date _____

Practice
Lesson 7

Half of a Half

Color to show the fraction.

1.

$\dfrac{1}{4}$

2.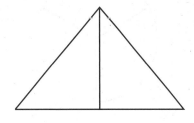

$\dfrac{1}{2}$

3.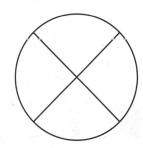

$\dfrac{3}{4}$

4.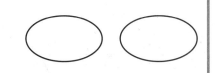

$\dfrac{1}{2}$

5.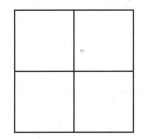

$\dfrac{2}{4}$

6.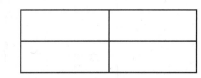

$\dfrac{1}{4}$

7.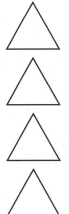

$\dfrac{3}{4}$

8.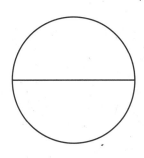

$\dfrac{1}{2}$

9.
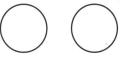

$\dfrac{2}{4}$

Chapter 8

Practice Book **P55**

Name _____ Date _____

Practice
Lesson 8

Thirds

Color to show the fraction.

1.

$\frac{2}{3}$

2.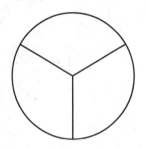

$\frac{1}{3}$

3.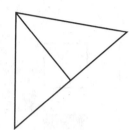

$\frac{1}{2}$

4.

$\frac{1}{3}$

5.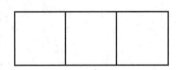

$\frac{3}{3}$

6.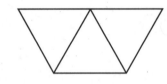

$\frac{2}{3}$

7.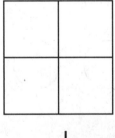

$\frac{1}{4}$

8.

$\frac{3}{4}$

9.

$\frac{1}{3}$

Name _____ Date _____

Practice
Lesson 1

Exploring Addition with Cuisenaire® Rods

Color the rods.
Complete each addition sentence.

If $\boxed{W}$ = 1, then...

1. Brown / W | Black
   $\boxed{1} + \boxed{7} = \boxed{8}$

2. Brown / Red | Dark Green
   $\boxed{} + \boxed{} = \boxed{8}$

3. Brown / Light Green | Yellow
   $\boxed{} + \boxed{} = \boxed{8}$

4. Brown / Purple | Purple
   $\boxed{} + \boxed{} = \boxed{8}$

Chapter 9

Name _____ Date _____

Practice
Lesson 2

Recording Addition Sentences

Color the rods.
Complete each addition sentence.

If $\boxed{W}$ = 1, then...

1. Black = Red + Yellow

   $\boxed{2} + \boxed{5} = \boxed{7}$

2. Dark Green = Red + Purple

   $\boxed{} + \boxed{} = \boxed{}$

3. Blue = Purple + Yellow

   $\boxed{} + \boxed{} = \boxed{}$

4. Orange = Yellow + Yellow

   $\boxed{} + \boxed{} = \boxed{}$

Exploring Input/Output Tables

What is missing?

1. subtract 3

in	out
9	6
5	2
10	
	4
	3
3	
	5

2. spend 10¢

in	out
53¢	43¢
67¢	
	68¢
70¢	
	83¢
	72¢
	90¢

3. 1 hour earlier

in	out
2:00	1:00
3:00	2:00
5:30	:
12:00	:
:	8:00
:	5:30
:	6:30

Chapter 9

Using Input/Output Tables

What is missing?

1.

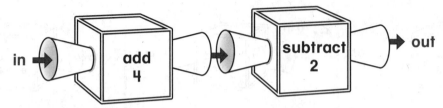

in	4	7	3	6	8	9	5
out	6	9	5	8	10	11	7

2.

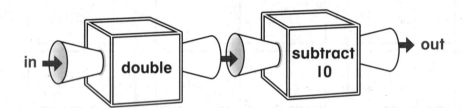

in	5	8	10	7	6	9	11
out	0	6	10	4	2	8	12

3.

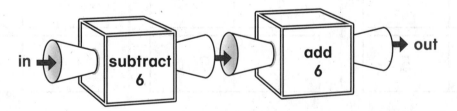

in	12	10	6	8	11	7	9
out	12	10	6	8	11	7	9

Name _____ Date _____

Practice Lesson 5

Making Fact Families

What is the fact family?

1.

   3 + 9 = 12

   ___ + ___ = ___

   12 − ___ = 9

   12 − ___ = ___

2.

   ___ + ___ = ___

   ___ + ___ = ___

   ___ − ___ = ___

   ___ − ___ = ___

3.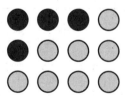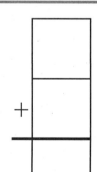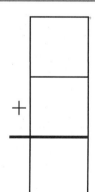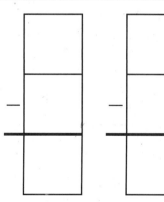

Chapter 9   Practice Book **P61**

Fact Families and Stair-Step Numbers

What is the fact family?

1.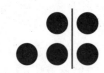

   3 + 2 = 5
   _____
   2 + 3 =
   _____
   5 − 2 =
   _____
   5 − 3 =
   _____

2.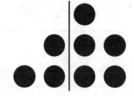

   3 + 5 =
   _____
   _ + _ =
   _____
   _ − _ =
   _____
   _ − _ =
   _____

3.

   2 + 4 =
   _____
   _ + _ =
   _____
   _ − _ =
   _____
   _ − _ =
   _____

4.

   2 + 5 =
   _____
   _ + _ =
   _____
   _ − _ =
   _____
   _ − _ =
   _____

5.

   8 + 1 =
   _____
   _ + _ =
   _____
   _ − _ =
   _____
   _ − _ =
   _____

Practice Lesson 7

Connecting Stories and Fact Families

What is the fact family?

1. Spot has 6 toys.
   2 are balls.
   The others are ropes.

   2 + 4 = 6
   4 + 2 = 6
   6 − 4 =
   6 − =

2. All of Jane's hats
   are red or blue.
   4 are red.
   5 are blue.

   ___ + ___ = ___
   ___ + ___ = ___
   ___ − ___ = ___
   ___ − ___ = ___

3. 3 fish are gold.
   The rest are blue.
   There are 11 fish.

   ___ + ___ = ___
   ___ + ___ = ___
   ___ − ___ = ___
   ___ − ___ = ___

Chapter 9

Practice Book **P63**

Name _____ Date _____

Practice
Lesson 8

Two-Sentence Fact Families

What is the fact family?

1.

___ + ___ = ___

___ + ___ = ___

___ − ___ = ___

___ − ___ = ___

2.

___ + ___ = ___

___ + ___ = ___

___ − ___ = ___

___ − ___ = ___

3. Liz has 6 flowers.
   3 are red.
   The rest are yellow.

   ___ + ___ = ___

   ___ − ___ = ___

4. Adam and Pete have the same number of marbles. Together they have 12 marbles.

   ___ + ___ = ___

   ___ − ___ = ___

Name _____ Date _____

Practice Lesson 1

Exploring Lines and Intersections

Draw dots to show where the lines intersect.

1.

2.

3.

Complete each table.

4.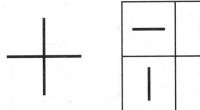

5.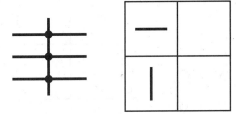

6.

7.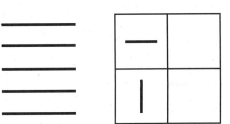

Chapter 10   Practice Book **P65**

Drawing Lines and Intersections

Draw dots to show where the lines intersect. Complete each table.

1.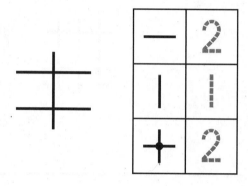

2.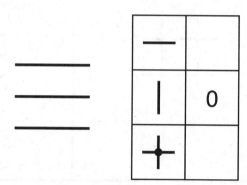

3.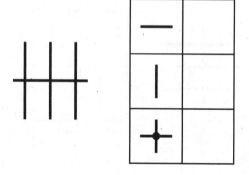

4.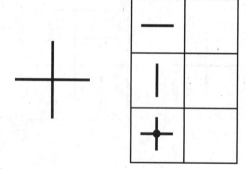

5.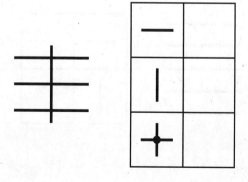

6.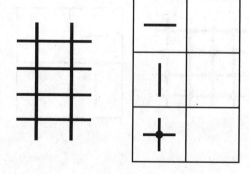

Name _____ Date _____

Practice Lesson 3

Exploring Direction on a Map

Find two paths from A to B on each map.

North ↑

1.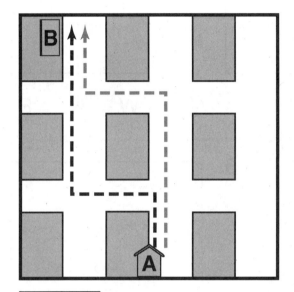

2.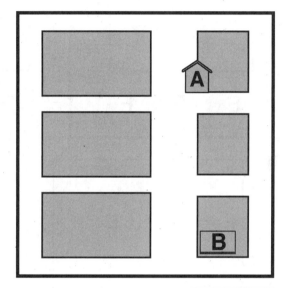

West ←

East →

3.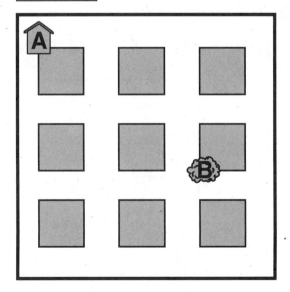

4.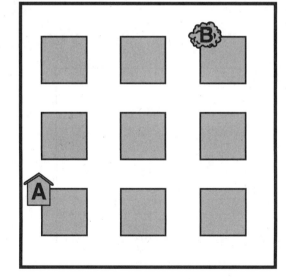

South ↓

Chapter 10

Practice Book **P67**

Name _____ Date _____

Practice
Lesson 4

Finding and Following Paths on a Grid

Start at the dot. Draw each path.

1.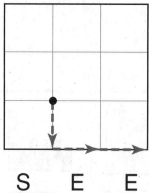
   S E E

2.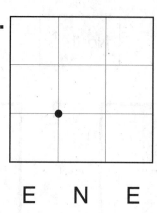
   E N E

3.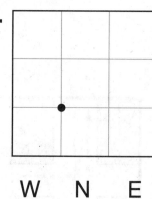
   W N E

4.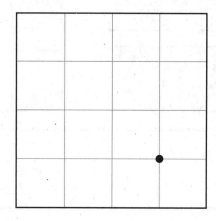
   N W N W

5.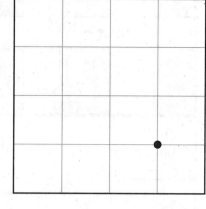
   S E N N

6.
   W S W N

7.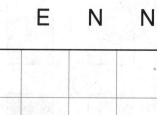
   W N N E

Practice
Lesson 5

Paths and Figures on a Grid

Start at the dot. Draw each path.

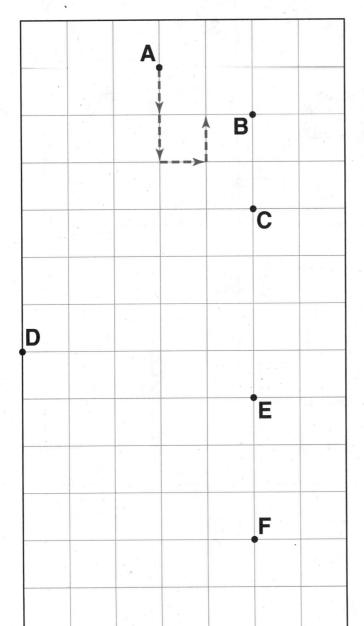

A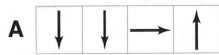

B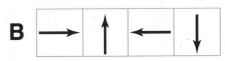

C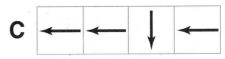

D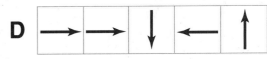

E ← ↑ → → ↓

F ↓ ← ← ↑ →

Chapter 10   Practice Book **P69**

Name _____   Date _____

Practice
Lesson 6

Exploring Symmetry

Draw the other half.

1.

2.

3.
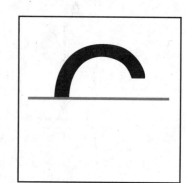

4.

5.
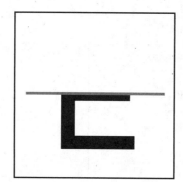

6.
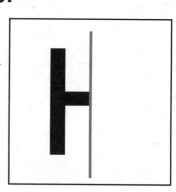

7.
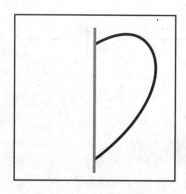

8.
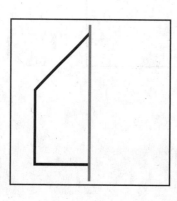

9.

P70 Practice Book

Chapter 10

Practice
Lesson 7

Connecting Points to Make Figures

H

A ---------- G

B•   •F

•C   •E

•D

1. Draw lines to connect the points.

   A and C   B and D   B and H   A and G
   G and E   H and F   D and F   C and E

2. How many triangles did you make?

   _____ triangles

3. How many squares did you make?

   _____ squares

Chapter 10   Practice Book **P71**

Name _____ Date _____

Practice Lesson 8

Investigating Rectangles

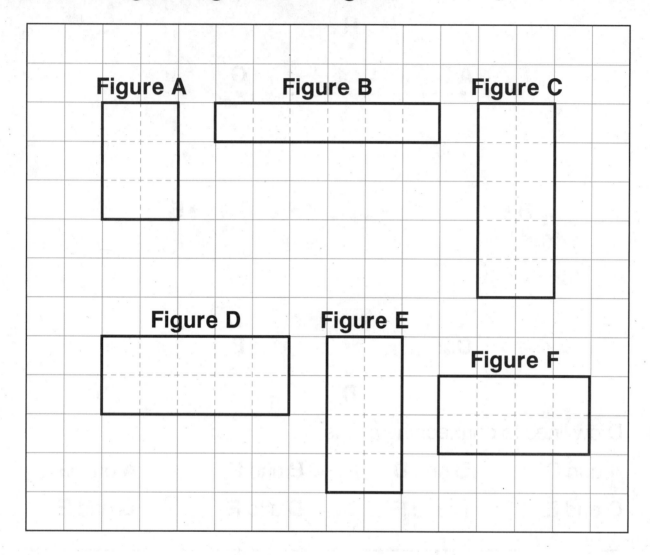

Use the figures above to complete the table.

Figure	A	B	C	D	E	F
Number of ☐	6					

Name _____ Date _____

Practice Lesson 9

Recording and Graphing Rectangles

1. How many ☐ are in each figure?

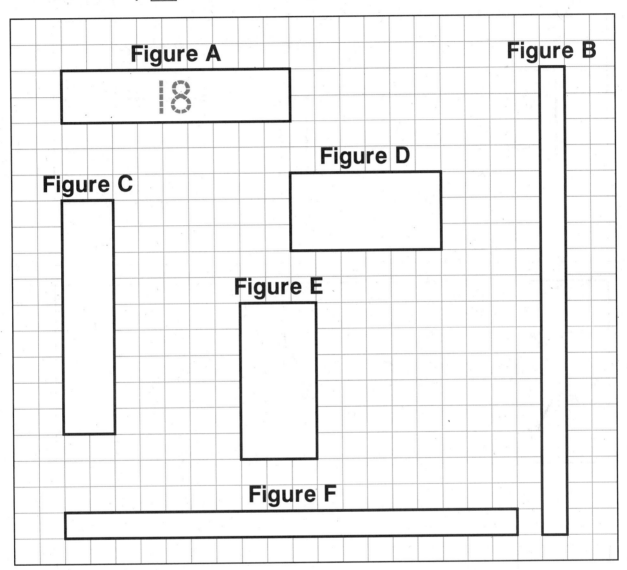

2. Use the figures above to complete the table.

Figure	A	B	C	D	E	F
Number of ☐	18					

Chapter 10

Practice Book **P73**

Name _____ Date _____

Practice
Lesson 10

Finding Congruent Figures

Which figure does not have the same size and shape?

1.

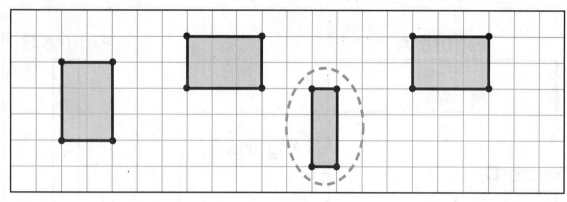

2.

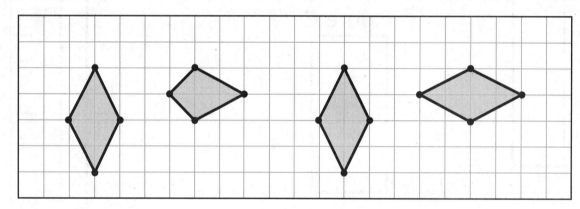

3.

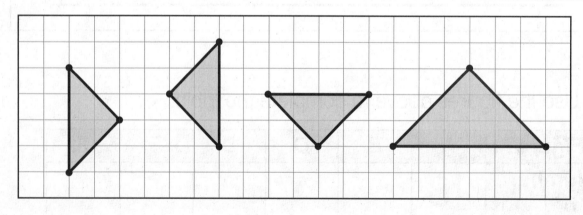

P74 Practice Book   Chapter 10

Name _____ Date _____

Practice Lesson 11

Exploring Three-Dimensional Figures

What is the shape of each object?

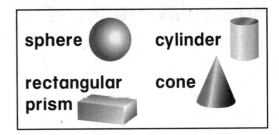

1.
sphere

2.
_____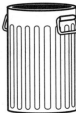

3.
_____

4.
_____

5.
_____

6.
_____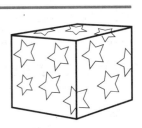

7. How are a cylinder and a sphere different?

_____

_____

Chapter 10

Practice Book **P75**

Comparing Groups

How many are there?
Write <, >, or =.

1.

2.

3.

4.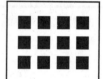

What is missing?

5.

6. 15 > ☐

7. 6 ◯ 8

8. 10 ◯ 10

9. 2 + 4 ◯ 2 + 7

10. 9 + 3 ◯ 3 + 5

P76 Practice Book

Chapter 11

Name _____ Date _____

Practice Lesson 2

Comparing Numbers and Temperatures

What numbers are missing?
Write <, >, or =.

1.
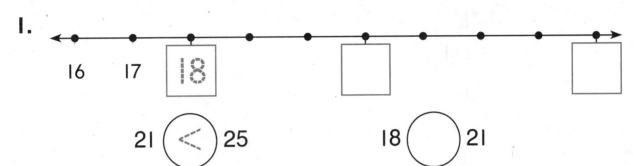

16 17 18 ☐ ☐

21 < 25   18 ◯ 21

2.
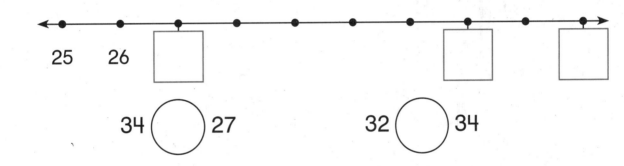

25 26 ☐   ☐ ☐

34 ◯ 27   32 ◯ 34

Write > or <.

3.
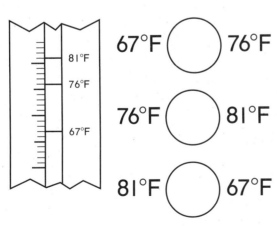

67°F ◯ 76°F

76°F ◯ 81°F

81°F ◯ 67°F

4.
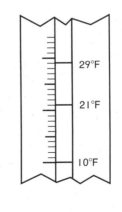

29°F ◯ 10°F

21°F ◯ 10°F

21°F ◯ 29°F

Chapter 11

Name _____ Date _____

Practice Lesson 3

Using Place Value to Compare Numbers

What numbers are shown?
Write <, >, or =.

1.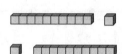

   [13] ◯ ☐

2.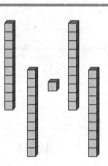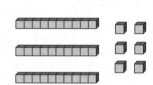

   ☐ ◯ ☐

Write >, <, or =.

3. 64 ◯ 64
4. 43 ◯ 47
5. 81 ◯ 18
6. 35 ◯ 53
7. 99 ◯ 99
8. 72 ◯ 27

P78 Practice Book   Chapter 11

Ordering Numbers

What is missing?
Write <, >, or =.

1.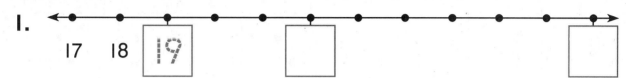

   17   18   19   ☐   ☐

   19 ◯ 28   28 ◯ 22   22 ◯ 19

2.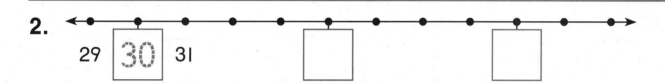

   29   30   31   ☐   ☐

   31 ◯ 30   34 ◯ 38   29 ◯ 34

Write the numbers.
Then order the numbers.

3.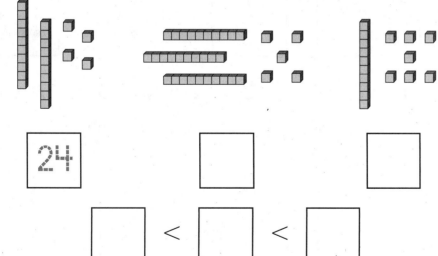

   24   ☐   ☐

   ☐ < ☐ < ☐

Chapter 11

Name _____ Date _____

Practice Lesson 5

Changing Both Sides of a Sentence

What is missing?

1.

3 4 and 3 ◯ 3 so 3 + 3 ◯ 4 + 3

2.
   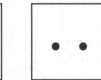

5 ◯ 4 and 2 ◯ 2 so 5 + 2 ◯ 4 + 2

3.

7 ◯ 9 and 2 ◯ 2 so 7 − 2 ◯ 9 − 2

4. Make your own.

☐ ◯ ☐ and ☐ ◯ ☐ so ☐ ◯ ☐

P80 Practice Book

Chapter 11

Name _____ Date _____

Practice Lesson 6

Comparing and Ordering Weights

apple napkin basketball cherries

Write <, >, or =.

1.

basketball ⟩ > ⟨ apple

2.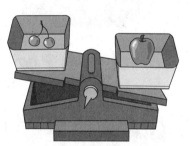

cherries ◯ apple

3.

napkin ◯ apple

4.

napkin ◯ cherries

5. Circle the lightest.

6. Circle the heaviest.

Chapter 11

Changing Both Pans of a Balance

What are the missing symbols and numbers?

1.

 and

6 oz ◯>◯ 2 oz   4 oz ◯ 4 oz

so   6 oz + 4 oz ◯ 2 oz + 4 oz

☐ oz ◯ ☐ oz

2.

 and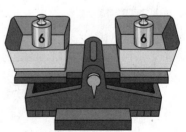

8 oz ◯ 12 oz   6 oz ◯ 6 oz

so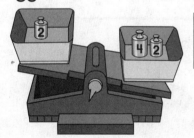

☐ oz − ☐ oz ◯ ☐ oz − ☐ oz

☐ oz ◯ ☐ oz

Name _____ Date _____

Practice
Lesson 1

Measuring Length with Nonstandard Units

**About how long is each line?
Use paper clips to measure.**

1. A ─────────────────────────────

   about _____ paper clips

2. B ──────────────

   about _____ paper clips

3. C ────────────────────

   about _____ paper clips

4. D ──────────────────────────

   about _____ paper clips

5. E ──────────────

   about _____ paper clips

6. Which line is the longest?

   _____

Chapter 12

Practice Book **P83**

Comparing and Ordering Lengths

1. Color the rods.

   orange
   blue
   brown
   black
   dark green
   yellow
   purple
   light green
   red
   w

Write >, <, or =.

2. red (<) blue

3. yellow ( ) black

4. red ( ) white

5. blue ( ) dark green

6. purple ( ) yellow

7. black ( ) brown

8. red + purple ( ) red + yellow

9. blue + black ( ) brown + blue

Name _____ Date _____

Practice
Lesson 3

Measuring with a Centimeter Ruler

Use a centimeter ruler. Draw each line.

1. 7 centimeters long

2. 10 centimeters long

3. a little more than 14 centimeters long

4. a little less than 5 centimeters long

5. Draw a square. Make each side 3 centimeters long.

Measuring with an Inch Ruler

Draw the lines.
Measure to the nearest inch.

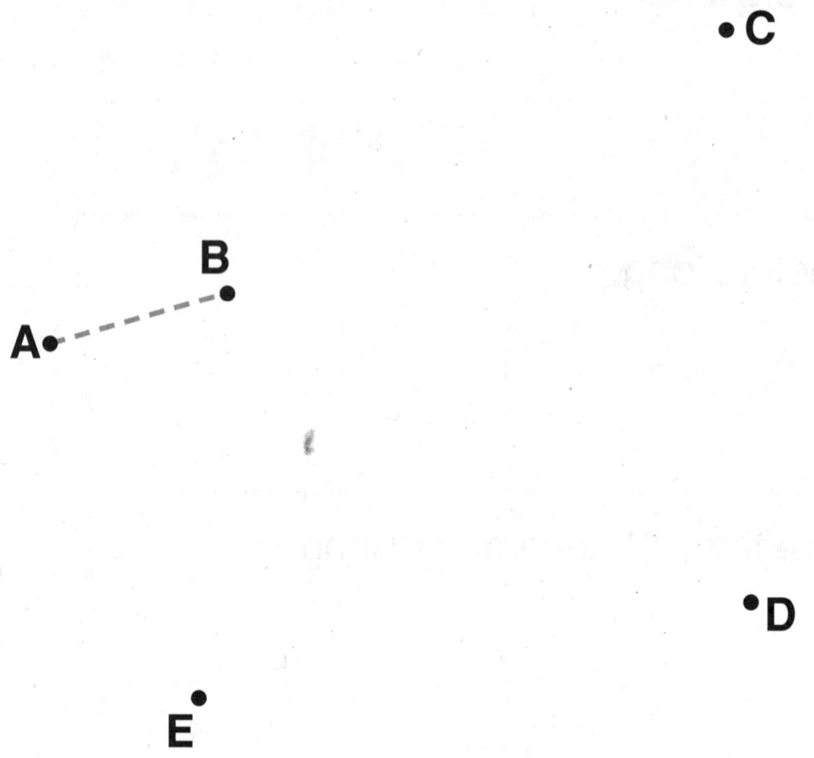

•C

B
•
A•- - -

•D

E
•

Line	Length
A to B	about ____ inch
B to C	about ____ inches
C to D	about ____ inches
D to E	about ____ inches
E to A	about ____ inches

Name _____ Date _____

Practice
Lesson 5

Comparing Figures by Size

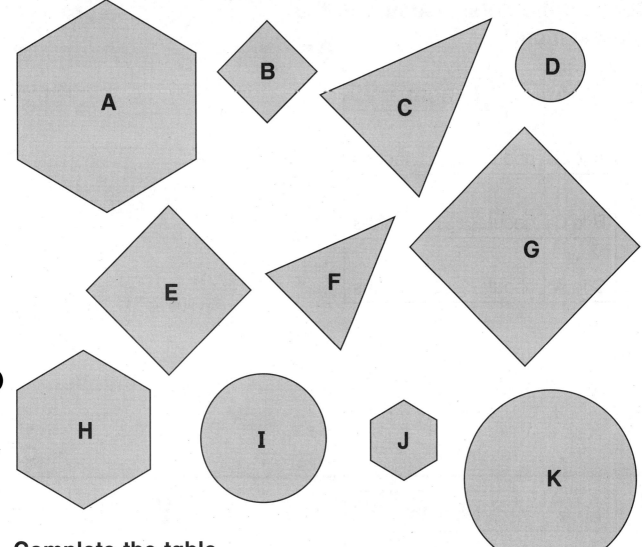

Complete the table.

	Largest	In-Between	Smallest
Square ▢	G		
Circle ◯		I	
Triangle △			
Hexagon ⬡			

Chapter 12

Practice Book **P87**

Name _____ Date _____

Practice
Lesson 6

Exploring Area

Connect the dots. Measure to the nearest inch.

1.

Line	Length
A to B	about ____4____ inches
B to C	about _____ inches
C to A	about _____ inches

2.

Line	Length
D to E	about _____ inches
E to F	about _____ inches
D to F	about _____ inches

3. Which triangle covers more area?

_____

Practice
Lesson 7

Finding Area on a Grid

What is the area of each shaded figure?

1.

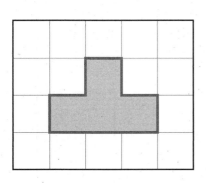

4

2.

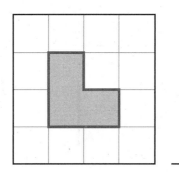

___

3.

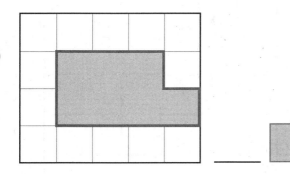

___

4.

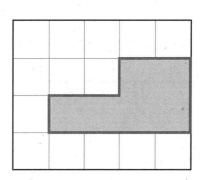

___

5.

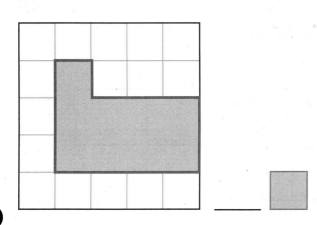

___

6.

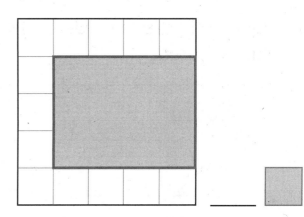

___

Chapter 12

Practice Book **P89**

Practice Lesson 8

Comparing Areas

1. What is the area of each figure?

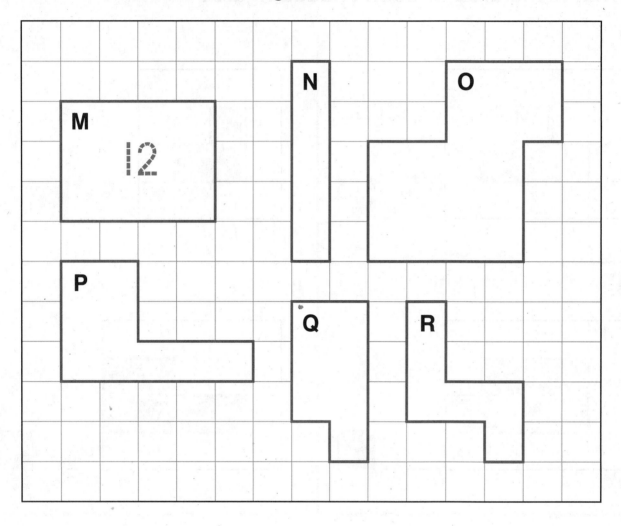

2. List the figures in order from smallest to largest area.

Figure	R				
Area	5				

Name _____ Date _____

Practice
Lesson 9

Measuring Boxes and Rectangles

What is the length of each side?

1.

about ____ cm

about 4 cm   A   about ____ cm

about ____ cm

2.

about ____ cm

about ____ cm   B   about ____ cm

about ____ cm

3. Will Rectangle A fit inside Rectangle B? _____

Will Rectangle B fit inside Rectangle A? _____

Chapter 12

Practice Book **P91**

Introducing Capacity with Nonstandard Units

Sean uses scoops to measure how much each container will hold.

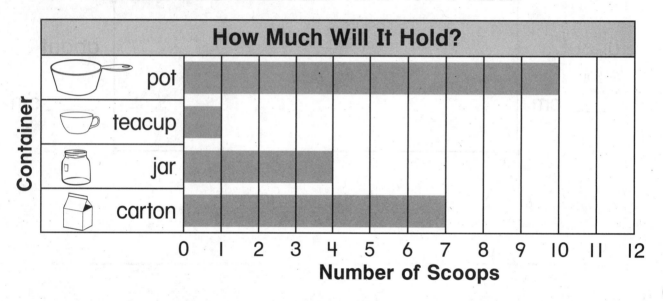

1. Which container holds the most? _____

2. How many more scoops does the carton hold than the jar? _____ more scoops

3. List the containers in order from holds the least to holds the most.

_____

_____

_____

Name _____ Date _____

Practice
Lesson 11

Measuring Capacity with Standard Units

Match to the best measurement in real life.

1. 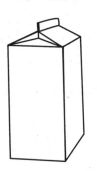   • 20 gallons

2.    • 30 inches

3.    • 1 pint

4.   • 6 inches

5.    • 2 quarts

Chapter 12

Name _____ Date _____

Practice
Lesson 1

Making Even and Odd Numbers

Draw and color rods to show each sum.
Is the sum even or odd?

1. 5 + 7

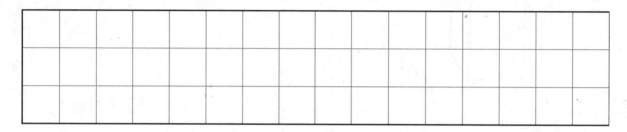

        even   odd

2. 6 + 3

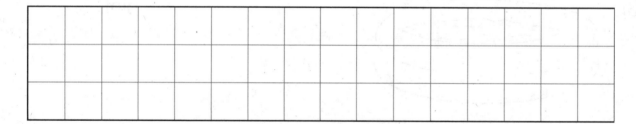

        even   odd

Draw a picture for each.
Is the sum even or odd?

3. odd + even = _____
4. odd + odd = _____

Name _____ Date _____

Practice Lesson 2

Making Numbers as Sums of 1, 2, 4, and 8

What number does each rod show?

1. purple
   4
   ____

2. W
   ____

3. brown
   ____

4. red
   ____

● What number sentence does each train show?

5. | brown | purple |

   ____ + ____ = ____

6. | brown | purple | red |

   ____ + ____ + ____ = ____

7. | brown | red | W |

   ____ + ____ + ____ = ____

Chapter 13

Practice Book **P95**

Combining Triangular Numbers

How many dots are in the triangle?

1.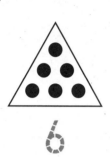

   6
   ___

2.

   ___

3.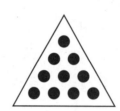

   ___

4.

   ___

What is the fact family?

5.

6.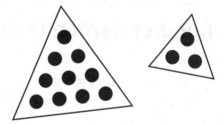

___ + ___ = ___   ___ + ___ = ___

___ + ___ = ___   ___ + ___ = ___

___ − ___ = ___   ___ − ___ = ___

___ − ___ = ___   ___ − ___ = ___

Name _____ Date _____

Practice
Lesson 4

Making Sums of 60

Complete each Cross Number Puzzle.
Use multiples of 10.

1.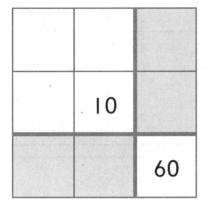

2.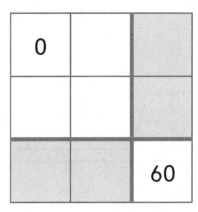

Continue the pattern.
Use the numbers to complete
the Cross Number Puzzles below.

5, 10, 15, __20__, ____, ____, ____, ____, ____, ____, ____, ____

3.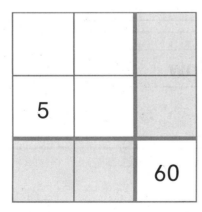

4.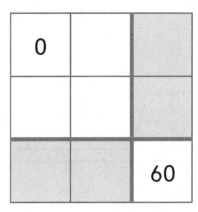

Chapter 13

Practice Book **P97**

Name _____ Date _____

Practice
Lesson 5

Sums to 12

Draw rods to show the sum. What is the sum?

1. 3 + 9 = ____

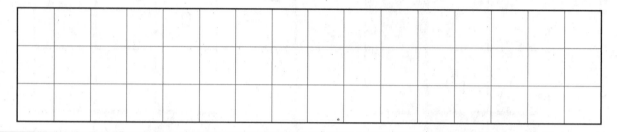

2. 4 + 7 = ____

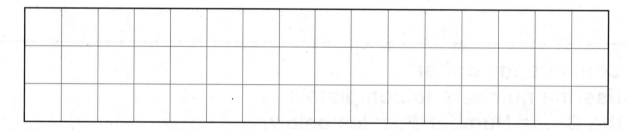

What number sentence does each train show?

3. | W | brown | W |

_____

4. | green | purple | yellow |

_____

5. | red | orange |

_____

P98 Practice Book Chapter 13

Name _____ Date _____

Practice
Lesson 6

Sums to 15

What number sentence does each train show?

1. | blue | yellow |

    _____

2. | black | brown |

    _____

Draw dots to show each addend.
Write the missing numbers.

3. 6 + 9

   [☐☐☐☐☐☐☐☐☐☐] [☐☐☐☐☐]

   10 + ___ = ___

4. 8 + 5

   [☐☐☐☐☐☐☐☐☐☐] [☐☐☐☐☐]

   10 + ___ = ___

Chapter 13

Practice Book **P99**

Sums to 16

Write the missing numbers.

1.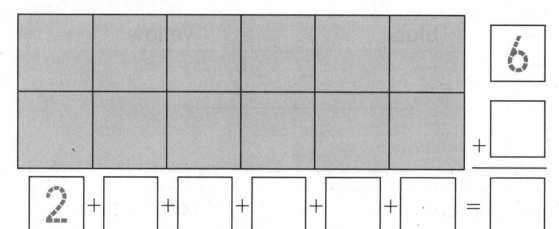

Complete each Cross Number Puzzle.

2.

	3	
1		6
		16

3.

4		12
6		14

4.

	2	
5		11
7		

5.

3		7
	7	
5		

Name _____ Date _____

Practice
Lesson 8

Sums to 18

Write the missing numbers.

1.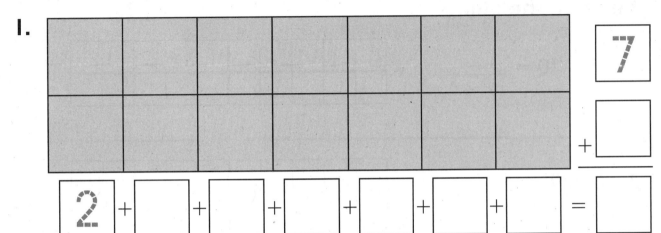

2. $5 + 6 =$ _____

   The sum is 1 more than △ + △ = _____.

   The sum is 1 less than ☐ + ☐ = _____.

3. $8 + 9 =$ _____

   The sum is 1 more than ◇ + ◇ = _____.

   The sum is 1 less than ⬡ + ⬡ = _____.

Practice Lesson 9

Sums to 20

Draw dots to show each number. Then find the sum

1. 9 + 10 = ____

2. 8 + 9 = ____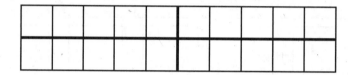

3. 7 + 5 = ____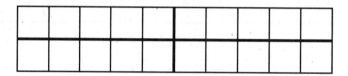

What is the sum?

4. 9 + 4 = ____	5. 10 + 10 = ____	6. 7 + 8 = ____
7. 7 + 5 = ____	8. 9 + 9 = ____	9. 7 + 7 = ____
10. 8 + 8 = ____	11. 6 + 7 = ____	12. 7 + 9 = ____

Adding Number Sentences

Write the missing numbers and signs.

1.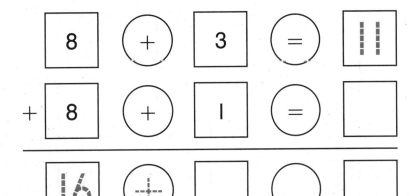

2.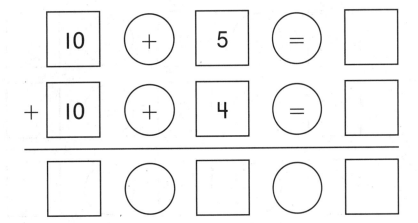

3.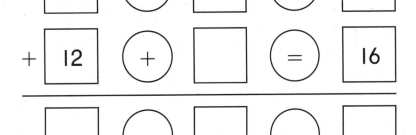

Name _____ Date _____

Practice Lesson 2

Making Addition Easier

Write the missing numbers and signs.

1.

2.

3.

P104 Practice Book

Chapter 14

Name _____ Date _____

Practice
Lesson 3

Modeling Number Sentences and Stories

Write the number sentence and answer the question.

Kirit built 3 birdhouses in 4 days.
Then he built 12 more birdhouses in 2 weeks.

1. How many birdhouses did Kirit build?

 + =

_____ birdhouses

2. How many days did it take?

 + =

_____ days

Katrina baked 12 muffins in 30 minutes.
Then she baked 18 more muffins in 1 hour.

3. How many muffins did Katrina bake?

 + =

_____ muffins

4. How many minutes did it take?

 + =

_____ minutes

Chapter 14

Practice Lesson 4

Making Subtraction Easier

Draw the jump.
Complete the number sentence.

1.
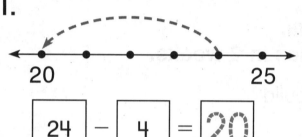

2.
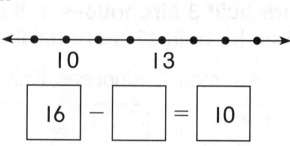

3.
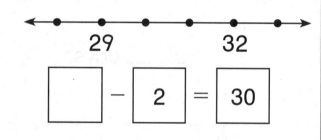

4.
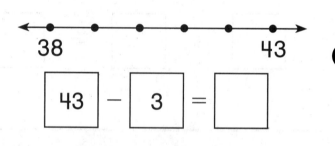

Write a number sentence and answer the question.

5. Marita had 68¢.
   She spent 1 dime on a whistle.
   How much money did she have left?

   ☐ − ☐ = ☐

   _____ ¢

Name _____ Date _____

Practice Lesson 5

Subtraction That Changes the Tens Digit

Draw the jump.
Complete the number sentence.

1.

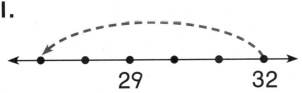

2.

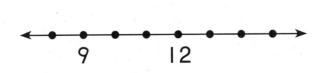

3.

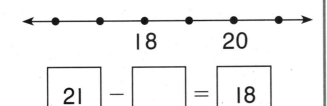

4.

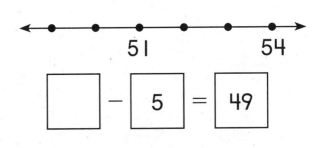

Write the number sentence and answer the question.

5. It was 65°F outside at noon.
   It was 6°F cooler after school.
   What was the temperature after school?

   ☐ − ☐ = ☐

   _____ °F

Chapter 14

Practice Book **P107**

Name _____ Date _____

Practice Lesson 6

Change from a Dollar

Write the missing amounts.
Use base-ten blocks to help.

1. Ryan spent this money.

_____7_____ ¢

He paid with a dollar bill. He got ___93___ ¢ change.

2. Mika spent this money.

_____ ¢

She paid with a dollar bill. She got _____ ¢ change.

3. This is my change from $1.

___15___ ¢

I spent _____ ¢.

4. This is my change from $1.

_____ ¢

I spent _____ ¢.

Name _____ Date _____

Practice
Lesson 7

Is This Story Reasonable?

**Solve. Is the story reasonable?
Circle *yes* or *no*.**

1. Josh made 14 peanut butter sandwiches and 15 cheese sandwiches for his lunch.

    Josh made _____ sandwiches for lunch.

    Is this story reasonable?   yes   no

2. Chita ate twice as many peaches as her brother. Her brother ate 2 peaches in one day.

    Chita ate _____ peaches in one day.

    Is this story reasonable?   yes   no

3. Write a reasonable story for 18 − 7.

    _____
    _____
    _____
    _____

Chapter 14

Practice Book **P109**

Name _____ Date _____

Practice Lesson 8

Solving Puzzles with Many Solutions

Solve each puzzle.

1.

20		
	30	50

2.

15		
	4	34

3. Find two ways to solve the puzzle.

		30
		25
10	45	

		30
		25
10	45	

P110 Practice Book — Chapter 14

Name _____ Date _____

Practice
Lesson 1

Identifying Rules

What is missing?
What is the rule?

1.

in	21	34	56	42		15
out	25	38	60		27	

The rule is _____.

2.

in	dog	cat	star	plant	
out	o	a	t		i

The rule is _____.

Chapter 15   Practice Book **P111**

Sorting Rules

What is missing?

1.

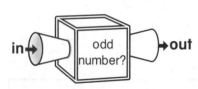

in	out
36	no
68	no
81	yes
	no
45	
	yes
72	

2.

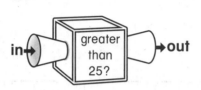

in	out
21	no
52	yes
73	yes
41	
17	
5	
	yes

3.

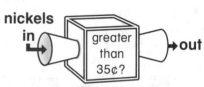

number of nickels	out
5	no
9	yes
8	yes
7	
3	
	no
10	

Name _____ Date _____

Practice
Lesson 3

Undoing Rules

Write the missing numbers.
What is the rule?

1.

in	11	25	52	84		46
out	22	36	63		76	

The rule is _____.

2.

in	15	39	27	92		61
out	4	28	16		73	

The rule is _____.

3. Do the rules undo each other? Explain.

_____

_____

_____

_____

Chapter 15

Practice Book **P113**

Name _____ Date _____

Practice
Lesson 4

Rules with More Than One Input

Write the missing numbers.

1.

in	21	70	55	41		75
in	32	16	44	27		
out	53	86			40	90

2.

in	1	36	72	8	26	
in	17	66	54	88	22	77
in	5	26	60	80		39
out	1	26	54		22	15

Name _____ Date _____

Practice
Lesson 5

Conversion Rules

Write the missing numbers.

1.

Number of quarts	1	2	3		6	7	
Number of cups	4	8	12	16			20

2.

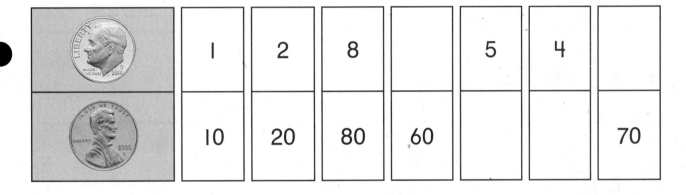

dime	1	2	8		5	4	
penny	10	20	80	60			70

3.

Number of quarts	1	6	3	7	4		
Number of pints	2	12	6			10	4

Chapter 15

Practice Book **P115**

Name _____ Date _____

Practice
Lesson 6

Skip-Counting with Money

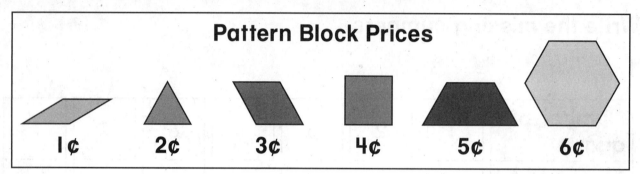

How much will the blocks cost?

1. 4 [trapezoid] _____¢

2. 3 [square] _____¢

3. 5 [rhombus] _____¢

4.  8 and 5 [triangle] _____¢

5. 2 and 4 [rhombus] _____¢

6. Emily bought half as many as [rhombus].
   She spent 12¢.
   How many did she buy?

   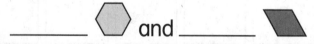 _____ and _____

7. Henry bought 5 blocks.
   They were all the same.
   They cost 25¢.
   Draw the type of block.

P116 Practice Book   Chapter 15

Name _____ Date _____

Practice
Lesson 7

Creating Figures and Patterns

Complete each table.

1. I am making snow globes.

Number of globes	1	2	3	4	5	6	7
Cost of ⬡	6¢	12¢	18¢				
Cost of ▱	5¢	10¢					
Total Cost	11¢	22¢					

2. I am making space ships.

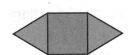

Number of rocket ships	1	2	4	6	7	9	10
Cost of ■	4¢						
Cost of ▲	2¢						
Cost of ▲	2¢						
Total Cost							

Chapter 15

Name _____ Date _____

Practice
Lesson 8

Patterns with Skip-Counting

Skip-count on the grids below.

1. Mark jumps of 5 with an X. Mark jumps of 9 with an O.

1	2	3	4	⨉5	6	7	8	ⓐ9	10
11	12	13	14	15	16	17	18	19	20
21	22	23	24	25	26	27	28	29	30
31	32	33	34	35	36	37	38	39	40
41	42	43	44	45	46	47	48	49	50

Where do the jumps meet? Find 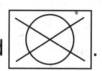. _____

2. Mark jumps of 8 with an X. Mark jumps of 4 with an O.

1	2	3	4	5	6	7	8	9	10
11	12	13	14	15	16	17	18	19	20
21	22	23	24	25	26	27	28	29	30
31	32	33	34	35	36	37	38	39	40

Where do the jumps meet? Find . _____

Practice
Lesson 9

● Relating One-Color Trains

Write the missing numbers.

1. **4** Red are as long as ____ Purple.

2. **2** Dark Green are as long as ____ Purple.

3. ____ Brown are as long as ____ Purple.

4. ____ Blue are as long as ____ Light Green or ____ Dark Green.

5. ____ Orange are as long as ____ Yellow.

Chapter 15